Setting Standards for Industry

Comparing the Emerging Chinese Standardization System and the Current US System

About the East-West Center

The East-West Center promotes better relations and understanding among the people and nations of the United States, Asia, and the Pacific through cooperative study, research, and dialogue. Established by the US Congress in 1960, the Center serves as a resource for information and analysis on critical issues of common concern, bringing people together to exchange views, build expertise, and develop policy options.

The Center's 21-acre Honolulu campus, adjacent to the University of Hawai'i at Mānoa, is located midway between Asia and the US mainland and features research, residential, and international conference facilities. The Center's Washington, DC, office focuses on preparing the United States for an era of growing Asia Pacific prominence.

The Center is an independent, public, nonprofit organization with funding from the US government, and additional support provided by private agencies, individuals, foundations, corporations, and governments in the region.

Policy Studies
an East-West Center series

Series Editors
Dieter Ernst and Marcus Mietzner

Description
Policy Studies presents original research on pressing economic and political policy challenges for governments and industry across Asia, and for the region's relations with the United States. Written for the policy and business communities, academics, journalists, and the informed public, the peer-reviewed publications in this series provide new policy insights and perspectives based on extensive fieldwork and rigorous scholarship.

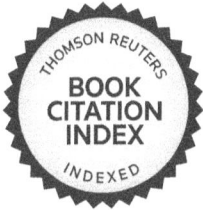

Policy Studies is indexed in the *Web of Science Book Citation Index*. The *Web of Science* is the largest and most comprehensive citation index available.

Notes to Contributors
Submissions may take the form of a proposal or complete manuscript. For more information on the Policy Studies series, please contact the Series Editors.

Editors, Policy Studies
East-West Center
1601 East-West Road
Honolulu, Hawai'i 96848-1601
Tel: 808.944.7197
Publications@EastWestCenter.org
EastWestCenter.org/PolicyStudies

Policy
Studies | 75

Setting Standards for Industry

Comparing the Emerging Chinese Standardization System and the Current US System

Liu Hui and Carl F. Cargill

Setting Standards for Industry: Comparing the Emerging Chinese Standardization System and the Current US System
Liu Hui and Carl F. Cargill

ISSN 1547-1349 (print) and 1547-1330 (electronic)
ISBN 978-0-86638-276-2 (print) and 978-0-86638-277-9 (electronic)

The views expressed are those of the author(s) and not necessarily those of the East-West Center.

Print copies are available from Amazon.com. Free electronic copies of most titles are available on the East-West Center website, at EastWestCenter.org/PolicyStudies, where submission guidelines can also be found. Questions about the series should be directed to:

Publications Office
East-West Center
1601 East-West Road
Honolulu, Hawai'i 96848-1601
Tel: 808.944.7145
Fax: 808.944.7376
EWCBooks@EastWestCenter.org
EastWestCenter.org/PolicyStudies

In Asia, print copies of all titles, and electronic copies of select Southeast Asia titles, co-published in Singapore, are available from:

ISEAS – Yusof Ishak Institute
30 Heng Mui Keng Terrace
Singapore 119614
pubsunit@iseas.edu.sg
bookshop.iseas.edu.sg

Contents

Summary

In every country, standardization is a reflection of that nation's level of industrialization. Creating consistent, widely adopted standards helps industries manufacture products in ways that are efficient, safe, repeatable, and of high quality. Standards are essential for translating new ideas, inventions, and discoveries into economic growth and prosperity. Whether standards originate from national governments, professional associations, private enterprises, or other standard-setting entities, they capture the interdependencies among the different sectors. On a broader level, they also embrace a nation's industrial, technical, and social policies. As nations change, standardization principles and practices change with them.

Nowhere is this more true than in China today, where a historic revision of the Chinese standardization regime is taking place. New methods, new ideas, and new strategies for effective standardization are percolating within Chinese government and civil society. One of the new ideas is "association standards," which are set by nonprofit, nongovernmental "social organizations" such as trade and professional associations. Though widespread in the United States, association standards are in the developmental stages in China, with reform efforts now focused on increasing the decision-making autonomy of nongovernmental standard-setting organizations.

This paper compares Chinese and American systems for setting industrial standards. Specifically, the paper compares the US system of voluntary standards, which relies on consensus among parties and market-driven initiatives, with current efforts to reform China's government-directed standardization system. The paper focuses on five aspects: the degree of development of these nonprofit associations, the abilities of the associations, government attitudes, market demand, and overseas experience. The paper culminates in a discussion of policy implications for China's reform efforts. An important argument is that the government should introduce pragmatic, feasible policy measures that address the needs and capabilities of standard-setting organizations. These policies can draw important lessons from the achievements of America's voluntary standard system. This would require a deep understanding of the advantages, disadvantages, and applicability of the US approach to voluntary standards.

Setting Standards for Industry
Comparing the Emerging Chinese Standardization System and the Current US System

Introduction

Standardization systems are unique to the country that they serve. Because standards represent an element of economic control, they should reflect the needs and aspirations of the government, society, business, and consumers. As governments and markets change, so should the standardization system. While China's government-led standardization management system reflected a model formed under the planned economy, key elements have failed to cope with the requirements of a market economy, and have thus hindered productivity.

China's standardization system has struggled to meet the demands of rapid economic and social development

In response, the Chinese standardization model is in the midst of change. Great progress has been made since the turn of the new century, but challenges remain. The standardization system has struggled to meet the demands of rapid economic and social development, occurring under pressure from changes in the domestic and foreign environment. At present, a series of problems exist, such as the unreasonable management and operation system, an unsound legal system for standardization, and an incomplete standard-setting and maintenance system, all of which compromise the effectiveness of standardization. In addition, many serious problems in China—such as nonexistent, ossified, or outdated standards and a single supply channel for standards—are, in fact, rooted in the standardization system.

China is now revising its Standardization Law, which was originally issued in 1988 as China began transitioning from a planned economy to a market economy and undergoing a general economic system reform. However, with the improvement of China's market economy, the advances in science and technology, and the deep changes in the function and model of government administration, the 1988 Standardization Law now in use has failed to meet the requirements of the current situation. One important goal of the latest revision of the Standardization Law is to provide legal support for standardization reform. To ensure a legal basis for major reforms and to align legislation to reform decisions, speeding up the revision of the 1988 law is imperative.

In recent years, particularly since 2013, China has quickened the pace of standardization reform. The State Council has released a series of documents relating to deeper reform, which also marked the official launch of a new round of standardization reform. These documents serve as a support and response to the reform campaign proposed by the Chinese government at the 18th National Congress of the Communist Party of China, held in 2012. The documents include the Deepening Standardization Reform Scheme (hereinafter referred to as the reform scheme), a programmatic document of China's standardization reform that further clarifies the future goals and reform tasks. The most striking highlight of the reform scheme lies in its clear provisions for fostering and developing association standards. The reform scheme emphasizes the

> *The most striking highlight of the reform scheme lies in its clear provisions for developing association standards*

need for streamlining administrative and delegating power, specifically by lessening government power to drive market power. This should allow nongovernmental, nonprofit "social organizations" more freedom to respond to market conditions and requirements. Now growing in numbers and influence in China, social organizations include standard-setting organizations, industry associations, technical organizations, and chambers of commerce. The reform scheme allows them to set general product and service standards, to be chosen by the market free from government control.

A second rationale for the reform scheme is the need to drive innovation. Standards serve as bridges between developing innovations and the marketization and industrialization of those innovations. Letting market entities set association standards speeds up the dissemination of innovative technologies. This, in turn, helps build a market-oriented technological innovation system that is based on commercial enterprises and encourages industry–university research cooperation. Finally, the reform scheme should help industry meet diversified market needs. As individualized consumption and diversified markets have increased, the practice of solely relying on government-driven standardization has proven inadequate for meeting market demands. Shifting the focus to association standards, on the other hand, can effectively increase the supply of standards that are necessary for fueling economic growth.

The drive to expand association standards in China reflects current economic, policy, and social needs. A key document—the *Decision on Major Issues Concerning the Comprehensively Deepening Reform of the Third Plenary Session of the 18th CPC Central Committee*—points out that the market needs to play a decisive role in resource allocation, and identifies requirements for developing association standards that are set and executed voluntarily by the market and social organizations. In addition, the definition of standards as being the products of voluntary consultation matches the reality of association standards quite well, since association standards can fully reflect the interests of standard makers and users. Therefore, finding common ground among all parties involved becomes possible, as association standards are better able to alleviate the conflicts of interests that plague China's existing standard system. At the same time, developing association standards conforms to China's need for comprehensive and deeper reforms. China's Standardization Administration has long supported the exploration and practice of association standards. In addition, China has fostered a group of qualified association standards makers, who have the ability to set, release, and popularize association standards, as well as to steadily advance the development of China's association standards on the basis of domestic and foreign experiences.

Literature Review

History of the US standardization system

In a systematic study on the emergence and development of the US standardization system, R.J. Robert (1999) found that the United States has a comparatively distinct standard system. The majority of US standards, including basic standards such as metering standards, are made by about 400 nongovernmental organizations, with the US government providing only consultation and guidance. This is in contrast to the much more common practice around the world of having standards approved and released by the government or government-authorized organizations—a practice that makes sense given the profound effects of standards on productivity, public security, and international competitiveness. This advisory role of the US government can be traced to the development of the nation's standardization. In *Trade Associations in Law and Business* (Kirsh and Shapiro 1939), the authors explain how American industry associations contributed to the formation of its standard system, noting that the country had only disparate and decentralized standards before World War I. In order to support the development of the war effort, five associations established the American Engineering Standards Committee (AESC) in 1918, from which separate industry-led standard associations developed gradually.

In the United States, standardization activities are carried out entirely by nongovernmental organizations

In the United States, standardization activities were carried out entirely by nongovernmental organizations without federal guidance or direction. It was not until the 1990s that the US government issued several laws in succession, including the National Technology Transfer and Advancement Act of 1995 and the Federal Participation in the Development and Use of Voluntary Consensus Standards and in Conformity Assessment Activities, which strengthened governmental support for standard-related activities and effectively enhanced the supply of standards (Russell 2006). In 2000, the United States released its first version of a national standard strategy, which established the basic principle of voluntary standardization and focused on association standards independently developed by the market. The goal was to unify its domestic market and increase the international competitiveness of its standards (Ernst 2013). In 2004, the US Congress released the Standards Development Organization Advancement Act and granted some exemptions from legal liabilities to formal standard-developing organizations in terms of anti-monopoly law and other areas. These exemptions specifically protected the rights of standard-setting organizations and aroused their enthusiasm for accelerating output (Lundqvist 2014). This model, which focused on nongovernmental organizations aided by gradually strengthened governmental support and guidance, developed into the standard system the United States has today. Of course, controversies never stop. For example, supporters believe that this system is conducive to technological development, while opponents believe it prejudices public interests, especially in this era when international standardization activities are led and supported by governments and national interests. In order to promote market unification and facilitate international trade, the European Community, predecessor to the European Union (EU), began to explore and establish a united and coordinated intraregional standard system in the 1980s. It carried out an aggressive reform of the old standard system (Pelkmans 1987), and by 1993, a "tripod" of standardization systems had taken shape.

Overview of standardization regimes around the world

The use of the word "regime" here connotes an entity that has a structured governance, in contrast to the word "system," which does not necessarily imply a governance scheme. Walter Mattli and Tim Büthe (2003) compared the standard regimes of the United States and Europe. They pointed out that there are two types of standardization regimes, with one being the standard regime of the United States that is fragmented, market-driven, and highly competitive internally. The other is the standard regime of Europe, which is hierarchical, highly coordinated, publicly regulated, and government-funded.

Another scholar, Song Hualin (2009), studied the development and evolution of domestic and overseas legal systems for technical standards. He believes that the evolution of China's technical standards tended to contrast with that of Western developed countries. Technical standards of Western countries were primarily "nongovernmental standards" made by nongovernmental organizations, and the effectiveness of such standards was guaranteed by the market competition mechanism instead of the coercive force of government. Later, with enhanced social regulation, administrative institutions gradually adopted privately made health, safety, and product standards by way of legal provisions, mutual agreements, or indirect recognition. China, on the other hand, has witnessed a gradual transformation from a unitary, mandatory standard system to the coexistence of mandatory and recommendation standards. Currently, the role of the market in the formation of the standard system serves to gradually strengthen standards, which is a different path from that of Western countries.

Based on an empirical study of the American standardization management system, Liao Li and Cheng Hong (2013) pointed out that the US system is technology-driven and law-based, with co-governance as its core idea. As China continues to progress and change, the current government-dominated supply of standards in China will be hard-pressed to satisfy the rich and diversified social needs of standardization. Increasingly, the role and status of social organizations such as industry associations and enterprises should be highlighted. This would help in establishing a demand-oriented standardization management system that better reflects the voluntary properties of the market and society, as well as the "soft law" nature of standards. The market should be allowed to select standards that meet the innovations and developments of the times, and that reflect scientific and technological levels that satisfy market needs. Standards involving public interests, such as health, safety, and environmental protections, should remain under the control of the government, since these are, in economic terms, impure public goods. In cases of market-driven standards, the government should participate in standardization activities as a partner with the market and society.

Association standards versus voluntary standards

Liu Sanjiang and Liu Hui (2015) believed that a core problem of China's standardization system reform was the supply system for standards. The present government-led supply model that China has adopted can no longer properly satisfy the demands of rapid economic and social development. Efforts should be made to make full use of social and market resources, tap social and market vitality, and expand the channel of supply so that standards can better meet the requirements of an increasingly complex infrastructure. Since the government can never establish a monopoly on the supply of impure public goods—

China's reform requires the creation of standardization co-governance structures and relationships

which are restricted by rationality, capacity, and resources—it has to strengthen cooperation with private sectors, such as nonprofit organizations and enterprises, and carry out co-governance. The issue of standardization system reform involves, in essence, the optimal distribution of responsibilities among the government, society, and the market. Essentially, reform requires the creation of standardization co-governance structures and relationships.

Wang Ping and Liag Zheng (2013) have studied the way that standardization has evolved among institutions and alliances after the reform and opening up in China. They believe that China has begun the shift from standards made completely by the government to the coexistence of governmental and nongovernmental standardization. This is of great significance to China's market economy transformation and the improvement of its industrial competitiveness. Many scholars have written on this subject, and their work stirs spirited discussions. Kang Junsheng and Yan Shaoqing (2015), for example, believe that the ideal management model of association standards should be one led by the government, dominated by social organizations such as associations, and supported by technical organizations. He Ming (2014) believes, based on comparisons between Chinese and American institution standards, that China should allow only professional, authoritative, and nationally recognized social associations that are specialized in standardization to make voluntary standards.

He advocates that a mechanism be established for transforming institution standards to national standards. Liu Jin and Wang Yanlin (2012) argue that the development of China's market economy requires the establishment of voluntary institution standards, and that China should introduce institution standards into the standardization law according to internationally accepted practices, as well as convert industry standards in the current law to institution standards. Wang Xia and Lu Lili (2010) believe that the development and implementation of institution standards could improve the overall industrial standardization, as well as technological development and innovation of enterprises within these associations. They view institution standards as being professional, innovative, and advanced.

Dieter Ernst (2013) suggests that for the United States the key to success is a bottom-up, decentralized, informal, market-led approach to standardization. He argues that there are significant differences in the organization and governance of standardization processes in the United States. These differences reflect the unique characteristics of each country's economic institutions, their levels of development, their economic growth models, and their cultures and histories. Andrew Updegrove (2010) believes that there are four important reasons for the foundation of standards consortia: obtaining unique benefits, an absence of alternatives, support of an existing standard, and displacement of a market incumbent. Francis L. "Tex" Criqui (2004) writes that basic skills improved through training, definite positioning, clear common goals, and mutual trust based on team interests are the four important factors driving the high level of competitiveness in the American standard system. Taking China and the United States as examples, Liu Fei (2009) indicates that the true value of standards lies in good systems and strategies, an argument based on an analysis of similarities and differences in standardization activities in the two economies. Zhang Shuqing (2007) believes that the American bottom-up standard-setting model is popular among all parties in an industry and adheres to the principle that standards should reflect technological progress and market demand as soon as possible.

In short, scholars have analyzed standardization management systems in China and other countries and have presented a wide set of beliefs and positions. Based on these discussions, the differences between association and voluntary standards will be further explored.

Overview of Association Standards

Before carrying out the comparative study on Chinese association standards and American voluntary standards, we must define what standards and their attributes are. This seemingly simple task has, in fact, great impact on our understanding of Chinese and American standardization systems.

Are standards voluntary or mandatory?

First, we should sort out international definitions of standards. The most authoritative definition is provided by the International Organization for Standardization. Their *ISO/IEC Guide 2:2004* defines standards as "documents, established by consensus and approved by recognized bodies, that provide, for common and repeated use, rules, guidelines or characteristics for activities or their results, aimed at the achievement of the optimum degree of order in a given context." This definition has now been accepted and identically adopted by standardization organizations in most countries, such as the British Standards Institution (BSI), the German Institute for

> *The ISO definition of standards has now been accepted and adopted by most countries*

Standardization (DIN), the Association Française de Normalisation (AFNOR), the Japanese Industrial Standards Committee (JISC), and others.

The World Trade Organization's technical barriers to trade (TBT) agreement defines standards as "non-mandatory documents approved by recognized bodies to provide, for common and repeated use, rules, guidelines or characteristics for products or relevant processing and production methods. Standards may also include or specify terms, notations, packing marks or labeling requirements for products, processing or production methods."

China has adopted a modification of the definition in *ISO/IEC Guide 2:2004*. The national standard GB/T 20000.1-2014, *Guide for Standardization Part 1: Standardization and Related Activities— General Vocabulary*, defines standards as "documents, established by consensus through standardization activities according to procedures as stated, that provide, for common and repeated use, rules, guidelines, or characteristics for activities or their results."

In view of these definitions of standards, we consider that standards are essentially voluntary technical normative documents that provide rules, guidelines, or characteristics for various activities or results and are designed for common and repeated use. The voluntary nature of standards is mainly reflected in two aspects:

1. The standards are made by technical committees composed of different parties whose participation is on a voluntary basis.
2. After standards are created and published, they are selected, adopted, and used by multiple different parties on a voluntary basis.

The coerciveness of standards—i.e., the enforceability of standards—is not an inherent property of a standard, but comes from the law and, thus, is an indirect attribute. An analysis from the legal perspective shows that in accordance with *ISO/IEC Guide 2:2004*, enforceability of standards comes from two sources: first, compliance with a standard is directly included in general laws as a legal obligation; second, a certain regulation makes exclusive reference to a specific standard, thereby making all other standards or options other than this one unsuitable for the intent and purpose of the regulation. In other words, standards can be enforceable only after they are stipulated by general laws or exclusively referred to by regulations (Liu 2016).

What are association standards?

"Association standards" is actually a general term for standards made by social associations, and includes institution standards and alliance standards. According to China's national standard GB/T 20004.1-2016, *Social Organization Standardization—Part 1: Guidelines for Good Practice*, association standards refer to standards made by self-governance and released and voluntarily adopted by associations according to their own (or their creating organizations') standard-setting procedures. In this definition, there are two terms to be further defined: associations and standards. According to the interpretation of GB/T 20004.1-2016, associations refer to social organizations with corporate capacity, corresponding professional expertise, standardization ability, and organization and management ability, such as societies, institutions, chambers of commerce, federations, and industrial technology alliances. The definition of standards in GB/T 20000.1-2014 ("documents, established by consensus through standardization activities according to procedures as stated, that provide, for common and repeated use, rules, guidelines, or characteristics for activities or their results") has been adopted by GB/T 20004.1-2016.

Institution standards. Institution standards are a significant category of association standards. They are initiated by industry associations, professional associations, and societies and developed jointly by stakeholders. Organizations capable of creating standards and willing to adopt standards have an opportunity to participate in the development of institution standards. The creators of institution standards make their own decisions on how they will respond to perceived market needs. Additionally, since these are voluntary standards and are not

> *Institution standards can promptly reflect technological progress and market demand*

legally enforceable or binding, the creators of the standards can voluntarily decide whether to adopt these standards or not, depending upon how they see the market developing. Because the development and revision cycle of institution standards is short, because they are more stringent than national standards, and because they are developed from the bottom up and in a voluntary environment, they can promptly reflect technological progress and market demand.

Alliance standards. Alliance standards are another important category of association standards. They are independently developed and issued by alliances according to their standard-setting procedures and voluntarily adopted by society. Alliance standards may be set by two types of alliances, which differ dramatically in the creation of standards and intellectual property rights policies, and in their openness to participants.

The first type of alliance is among nonprofit organizations similar to institutions. They have open organizational structures and memberships and relatively reasonable standard-setting procedures. They differ from traditional standardization by using consensus, and they regard speed as one of their important principles. Standards set by these alliances should be considered as association (institution) standards, and can be either pure or impure public goods. This type of organization is also faced with patent problems, but it generally accepts the reasonable and nondiscriminatory principle (i.e., the RAND or FRAND principle).

A second type of alliance can more properly be described as interest groups consisting of a small number of leading enterprises and consortia that attempt to occupy the market with standards and patents (Wang and Liang 2016). These organizations, in general, involve very narrow expertise, set high entry thresholds, and do not use regular standard-setting procedures. Members of these organizations consist of a few leading enterprises in the industry, which join forces to control technical patents and incorporate patented technologies into standards.

> *Alliance standards are aimed ultimately at monopoly of the market by a few enterprises*

This control facilitates cross-licensing among members and allows them to charge high patent fees to nonmembers who need to implement the standards. Standards set by these alliances are not public, but are strongly private goods, aimed ultimately at monopoly of the market by a few enterprises through a combination of patent rights and standards.

Motives of association standardization

Before studying association standards, we need to understand why association standardization exists and why social organizations set association standards. In China, the setting of association standards is remarkably different from the setting of national, industrial, and provincial standards. National, industrial, and provincial standards are made by the government and consist of recommendation and mandatory standards; the setting of association standards is a form of spontaneous, market-driven, bottom-up behavior.

Association standardization is usually driven by two motives. The first is aimed at normalizing enterprise production and strengthening industry self-discipline. Alliance standards are set to systematize enterprise production, eliminate production without standards, and guide healthy and orderly development of enterprises or industrial clusters. For this purpose, alliance standards are set in many industries or industrial clusters in China, especially enterprises or clusters characterized by low technical content in products, small scale, and weak innovation ability. These standards are most typical in the Pearl River Delta and Yangtze River Delta regions in China. The alliance standard for rosewood furniture in Guangdong Province is an excellent example. Alliance standards in Zhongshan, Guangdong Province, originated from Da Yong Rosewood Furniture, the first industry to set alliance standards in Zhongshan. The industry dated from the late 1970s, when there were mostly family workshops with low technical capacity due to the absence of production standards and a low entry threshold. With fierce market competition driven by cost, some enterprises reduced costs by using cheap raw materials and jerry-building. This hindered the healthy development of the industry throughout China and seriously restricted the development and growth of Da Yong Rosewood Furniture. The situation was typical of numerous industrial clusters in Zhongshan. Disordered competition disrupted the healthy growth of the market and forced the creation of "standards." In order to solve the development bottleneck of companies such as Da Yong Rosewood Furniture, Zhongshan Rosewood Furniture Association, in coordination with the local government, made an alliance standard of "valuable hardwood furniture in deep color" for their members; this later became an industry standard. The implementation of the alliance standard led to a dramatic drop in the number of rosewood furniture enterprises, from more than 400 at the peak to about 170 at present. However, this drop was more than countered by a rapid increase in production value, from several hundred million yuan to nearly two billion yuan. The standard allowed members of the Zhongshan Rosewood Furniture Association to increase their share of the nationwide rosewood furniture market to more than 60 percent, becoming China's largest rosewood furniture production base (Shang 2014). Other examples include the alliance standard for padlocks in Jinhua, Zhejiang Province; the one for quartz artware in Pujiang, Zhejiang Province; and the one for scissors in Tuorong County, Ningde, Fujian Province.

A second motive for association standardization is to respond rapidly to technological changes in a certain industry or industrial subset and to achieve competitive advantages via technology application. Take alliance standards in the information and communications technology (ICT) industry as an example. The ICT industry constantly experiences the emergence of new technologies, products, and services, but since current standard-developing organizations lack technical capabilities to set new standards, the demand for rapid technological development cannot be satisfied (Updegrove 2010). In this case, leading technology enterprises or enterprise groups have increased market share of products and relevant technologies through market expansion, growing until they occupy the main markets and their technical standards become de facto standards. They have then promoted the application of these standards by establishing patent alliances (Yao and Song 2010), such as DVD 6C. The 6C alliance was founded by six enterprises: Hitachi, Panasonic, Mitsubishi, Time Warner, Toshiba, and JVC, all of which have the essential standard-required patents. In the process of DVD standard setting and implementation, leading technological enterprises always play a dominant role, and full competition enables new DVD technology standards to be well adapted to the dynamic changes in market demand and technology. On June 11, 1999, the 6C alliance released a joint statement on DVD patent licensing, which initiated joint licensing for core patents concerning DVD specifications applied to video players, recorders, drivers, video discs, and recordable compact discs owned by the six members worldwide.

Relationship between voluntary standards and association standards

The concept of voluntary standards is relative to mandatory standards. Standards are described by technical barriers to trade (TBT) agreements as voluntary, so the term "voluntary standard," indicating voluntary participation in standards setting and adoption, applies. The standard systems of most countries in the world are voluntary. With the US standard system as an example, not only are institution standards and alliance standards voluntary, but most of America's national standards are also voluntary. This indicates that association standards are mostly voluntary as well, since most national standards are also voluntary. Therefore, association standards are defined mainly by the applicability of the standards, which is limited to within associations or alliances. Actually, comparison of the two can only be made as to whether standards are voluntary or mandatory; with respect to the range of application of standards, voluntary standards can be applied throughout a country, an industry, an association or alliance, and even an enterprise.

Current State of Chinese and American Association and Voluntary Standards

The Chinese standardization landscape

In China, standards at different national levels are now set in different ways, depending upon the formality of the process and the aims of the government. National standards are set by the Standardization Administration of China (SAC). The SAC authorizes its subordinate technical committees to set standards for different fields. Industrial standards are set by the technical committees of respective industries and are authorized by competent authorities. Provincial standards are set by local technical committees, which are authorized by local competent administrative departments in charge of standardization. These standards reflect government needs and the national interest.

However, since the 1980s and 1990s, a new type of standard that exceeds the current standard system has emerged. These are association standards (mentioned earlier), which are set by social organizations—specifically, public societies and associations—and by industrial alliances in China in response to market needs. At present, social organizations undertaking standardization work in China are mostly industry associations or government-run nongovernmental organizations (Wang and Liang 2016). The

Since the 1980s and 1990s, a new type of standard has emerged in China: association standards

creation of these organizations is highly encouraged by the government. The rationale for this lies in the fact that national and industrial standards are difficult to create, and the ability to generate large numbers of them annually is limited. Additionally, in regards to setting national and industrial standards, very few applications from nongovernmental standardization organizations are approved each year. As a result, many standardization organizations, especially nongovernmental standardization organizations, prefer to create and release association standards. As the country gives a freer rein to association standards, China will see their numbers increase.

As it stands, China's association standards mainly include institution standards and, more commonly, alliance standards. Although association standards are neither included in the current formal standardization system nor given a civil administrative enforcement mechanism, these standards are building momentum in China. They are appearing in an increasing number of sectors and constantly expanding their geographic reach, as more and more are being effectively created and used.

Characteristics of association standardization. As noted above, China's association standards have developed mainly from alliance standards. Currently, three characteristics define the country's association standardization efforts. First, there is a continual geographical expansion of association standards. Association standards originated from the developed Pearl River Delta and Yangtze River Delta regions and gradually expanded nationwide. Among the association standards, alliance standards have undergone the most rapid development. The growth began in 1998 when more than 10 enterprises in Da Yong Town, Zhongshan, Guangdong Province, united to develop and implement the rosewood furniture standards, which became one of the earliest alliance standards. In 2005, Guangdong Vanward Group Co., Ltd. and Guangdong Macro Gas Appliance Co., Ltd. in Shunde District, Foshan, Guangdong Province, jointly developed and implemented the alliance standard of "condensing domestic gas instantaneous water heaters," marking the advent of the first domestic regional alliance standard. Similar alliance standards appeared very quickly in developed southeastern coastal provinces such as Guangdong, Zhejiang, Fujian, and Shandong.

Association standards are seen not only in high-tech industries, but also in traditional industries

Next, there is a continual extension of association standards into different sectors. Association standards are seen not only in high-tech industries with rapid technological updates and short product cycles, but are also occurring in a range of competitive traditional industries with mature technology and slowing development. For example, the intelligent grouping and resource-sharing protocol (IGRS) in the electronic information sector, the LED industry alliance standard in the semiconductor lighting sector, and the ultra–high frequency radio frequency identification alliance standard in the communication sector are all association standards developed and implemented by high-tech enterprises mastering core technologies in response to market-driven technical changes in their sectors.

Finally, there is continual growth in the adoption of association standards. Despite the absence of general statistics, it is clear that the presence of a single association standard in an industry, with certification marks applied to products meeting that standard, has attracted more voluntary users and promoted the adoption of standards in many cases. For example, the domestic solar water-heating system standard set by the Haining Solar Energy Industries Association of Zhejiang Province (which combined a special mark and an anti-counterfeiting inquiry system) has not only been effectively implemented by member enterprises, but was also voluntarily taken up by 10 percent of nonmember enterprises after its release. In addition, the number of enterprises using the alliance standard for industrial washing machines in Panyu, Guangdong, has climbed by 200 percent since the standard was published.

In 2015, SAC launched an experimental project for association standardization nationwide. A total of 39 pilot associations, including the Chinese Institute of Electronics and the China Association of Chinese Medicine, were in the first group of associations chosen. At present, the National Information Platform for Association Standards has been created, and 225 association organizations have registered to carry forward their standardization programs.

Provisions on association standards involving patents. On December 19, 2013, the SAC and the State Intellectual Property Office jointly released the interim "Regulation Measures on National Standards Involving Patents" (hereinafter referred to as the regulation).

It points out that patents involved in national standards shall be essential patents. In respect of disclosure of patent information, the regulation stipulates that in any stage of the standard formulation or revision process, organizations or individuals participating in the formulation or revision of standards shall disclose, in a timely manner, essential patents that they know about and/or possess. Organizations or individuals participating in standards formulation shall assume the corresponding liability for failure to disclose patents they possess as required. The SAC shall publicize the full text of draft standards and known patent information for a period of 30 days before national standards that involve or might involve patents are approved and released.

In respect of patent licensing, the regulation points out that where national standards involve any patents in the process of development and revision, the SAC or competent authorities shall require patentees or patent applicants to make a patent licensing declaration in a timely manner. Such declarations shall include one option selected by the patentee/patent applicant from the following three options:

1. The patentee/patent applicant is willing to license any organization or individual, free of charge, and on a reasonable and nondiscriminatory basis, to practice his/her patent when implementing the national standard.
2. The patentee/patent applicant is willing to license any organization or individual, on a reasonable and nondiscriminatory basis, to practice his/her patent when implementing the national standard.
3. The patentee or patent applicant is not willing to license pursuant to either one of the aforesaid options.

Except for mandatory national standards, where a patent license granted by patentees or patent applicants as per first or second provisions above is not obtained, national standards shall not contain articles based on such patents. Drafts of national standards under the above-mentioned circumstance shall not be approved or published. After a national standard is published, if it is found to involve patents but have no patent license, patent-licensing declarations shall be obtained from the patentees or patent applicants within the stipulated time. Except for mandatory national standards, in case of failure to obtain a patent license granted by patentees or patent applicants as per the first or second provisions set forth above within the prescribed time frame, the SAC may suspend the implementation of corresponding national standards.

With respect to licensing fees, the regulation points out that patent licensing and licensing fees involving national standards shall be agreed upon consultation by standard users and patentees or patent applicants according to patent licensing declarations made by the latter.

Current state of American voluntary standards

Unlike other countries, standard-setting organizations in the United States first appeared in private sectors, and were aimed at satisfying the special requirements of these sectors and solving problems in production and engineering. Early pioneers of standardization in the United States were science and technology societies. These societies included the American Society of Civil Engineers (ASCE) founded in 1852, the American Society of Mechanical Engineers (ASME)

> *Unlike other countries, US standard-setting organizations first appeared in private sectors*

founded in 1880, and the American Society for Testing and Materials (ASTM) founded in 1898, as well as trade associations such as the American Iron and Steel Institute (AISI) founded in 1855. From the very beginning, these societies and institutes have had the right to set the standards employed by US industries.

The American voluntary standard system consists mainly of American National Standards, society (institution) standards, and alliance standards. To clarify, American National Standards refer to those developed by accredited standard committees under the aegis of the American National Standards Institute (ANSI). Society standards are those developed by accredited organizations such as the Institute of Electrical and Electronics Engineers (IEEE), also operating under ANSI rules. Alliance standards refer to consortia. Voluntary standards feature voluntary participation in standard setting and voluntary application. American society (institution) standards are set by various society (institution) organizations, where all directly affected producers, users, and consumers, including the government and academia, can participate. Alliances, unlike societies (institutions), are committed to rapidly setting standards that can reflect the latest technology in industries and are voluntarily formed by several enterprises to achieve common interests. Alliance standards are normative documents set by alliance members upon consultation to satisfy their own needs.

American voluntary standards are set mainly by nongovernmental organizations. These nongovernmental organizations (mostly industry associations and professional organizations) can set and release standards regarding their own professions or industries, and approve release of institution standards with respective identifying numbers and nomenclature. Therefore, American standards are decentralized

American standards are decentralized and diversified

and diversified. The American National Standards Institute (ANSI), the management and coordination agency for the American voluntary standard system excluding consortia, does not create standards. Instead, it adopts as national standards those basic private standards that have national influence, and it grants those that have been created under ANSI rules with an ANSI code. (Consortia, which do not participate under ANSI rules or requirements, do not receive the American national standard designation, granted solely by ANSI). Thus, American national and institution standards are underpinned by nongovernmental organizations.

Comparative Analysis of Factors Influencing Chinese and American Association Standards

At present, there is no specific research on the factors that have shaped the development of association standards. However, from a general perspective, two considerations influence how association standards develop: internal factors and external factors. The internal influencing factors include: (1) the strength of nonprofit organizations and (2) the capability of associations to carry out standardization. The external influence factors include: (1) the government's attitude, (2) market demand, and (3) relevant foreign experience. These five factors have had varying impacts on the way that association standards developed in China and the United States.

The strength of nonprofit organizations

In China, association standard–setting organizations are mainly nonprofit organizations and alliances of special interests, including various associations, institutes, chambers of commerce, unions, and industrial technology alliances. Only a small number of China's existing industry associations have set up departments responsible for standardization. These associations can be divided into three categories (Wang and Lu 2010):

1. **Industry associations**, such as the China Machinery Industry Federation with a standardization department; the China Electricity Council with a standardization center; and the China Chamber of Commerce with the Department of Industry Development responsible for standardization. Such association standardization departments are in charge of the organization and plan approval and advertising of industry-wide standardization.
2. **Professional associations**, such as the China Electronic Commerce Association with an electronic commerce application center as its standardization department; the China Automobile Industry Association with the Department of Industry Development responsible for standardization; and the China National Coal Association with the Department of Science and Technology Development. These association standardization departments are only responsible for setting standards in their specialized fields.
3. **Technical committees** of industries subordinated to associations. No additional standardization department is set up for such associations, and the technical committee is responsible for standard setting in the industry, such as the China Communications Standards Association, China Battery Industry Association, and China Association of Lighting Industry.

In the United States, nonprofit organizations represent a large and diverse sector that plays an important role in the country's society, economy, and politics, especially in providing services, advocacy, expression, and community building (Holland and Ritvo 2008). As of

2015, 1.57 million nonprofit organizations were registered with the US Internal Revenue Service, including 1.09 million public charitable organizations and more than 100,000 private foundations, as well as over 450,000 nonprofit organizations of other types, such as chambers of commerce.[1]

Standardization organizations in the United States mainly include professional groups and consortia engaged in standardization activities. Among them, professional groups can be broadly divided into three categories. First are standardization bodies, such as nongovernmental organizations dedicated to standard setting, also known as accredited standards committees under ANSI nomenclature. Second are professional scientific associations, which now number over two thousand. These are academic organizations consisting of scientists, engineers, and technical personnel and are formed to conduct academic exchanges. They include some institutes, such as the Institute of Electrical and Electronics Engineers (IEEE), known as accredited organizations under ANSI nomenclature when they develop standards. Third are industry and trade associations that are voluntarily organized by small and large manufacturers to provide information services to members and to develop product standards designed to develop the industry, promote trade, and increase profits. Examples include the famous Aerospace Industry Association (AIA) and the American Petroleum Institute (API), which are known as accredited organizations under ANSI nomenclature.

Another type of standardization organization in the United States is the consortium. Generally, special groups form a consortium to set standards, instead of following the traditional work procedures under ANSI rules. Each consortium has its own rules and procedures for developing consortia specifications, which are used as standards within the industry.

China lags behind the United States in terms of the maturity of its nongovernmental organizations for two primary reasons. One is that very few private standardization organizations exist, since association standardization has just started in China. At present, the vast majority of the 225 associations registered in the information platform for national association standards are government-backed. In the United States, there are 295 accredited standard developers accepted by the ANSI, the majority of which are self-governed bodies. In addition, throughout the United States there are more than 700 standard-setting organizations, which include consortia.

The second reason for China's low number of nongovernment standardization organizations is simply its development history—that is, the short amount of time these groups have had to grow. Most of China's association standardization organizations were established after 1978, when reform and opening up began, and only a few institutes and associations were created in the nearly three decades spanning 1978 and the establishment of new China in 1949. For example, the Chinese Hydraulic Engineering Society, formerly known as the Chinese Institute of Hydraulic Engineering, was founded in 1931 and renamed in 1957. The Chinese Society of Electrical Engineering, formerly known as the Chinese Institute of Electrical Engineers, was founded in 1934 and renamed in 1958. The Chinese Mechanical Engineering Society was founded in 1936. Many US associations have longer histories than those of China, such as the American Society for Testing and Materials (ASTM), the American Society of Mechanical Engineers (ASME), and UL (formerly Underwriters Laboratories), which were established in the late nineteenth century.

> *Most of China's association standardization organizations were established after 1978 when reform began*

Additionally, another significant difference between nonprofit organizations in China and the United States is that China's existing nonprofit organizations include many institutes, associations, chambers of commerce, unions, and industrial technology alliances that have a government background. These organizations are associated with and, to a certain extent, products of government reform and dependent on government support. Their main resources—including human, financial, material, informational, management, and organizational resources—are mainly provided by powerful, monopolistic government agencies. Therefore, these organizations adopt an administrative, top-down bureaucracy simulating the government (Wang and Jia 2003).

Factors influencing an organization's standardization capability

The standardization capability of nonprofit organizations is driven by multiple considerations. These include the size and reputation of an organization, the ability of the organization to grow or scale, and the revenue that the organization receives in order to grow or maintain itself. The ability of the organization to prosper is also driven by how much it receives in research and development (R&D) input to feed the creation of standards, how well it manages and maintains the quality of its standards, how much intellectual property is at stake, and other considerations. Finally, the ability of the organization's staff to provide administrative support can make or break any standardization effort. All of these factors should be considered when evaluating the capabilities of an organization to successfully create standards.

However, Zeng Deming et al. (2005) have proposed different criteria. They considered that technical superiority, technical standard-setting ability, and standard-promotion ability are key indicators for measuring the standardization capability of organizations. The technical superiority of the organization (its members and undertakings) provides a fundamental guarantee that the standards set by that organization will help obtain the underlying economic interests. This has a powerful effect on the standard-setting process. At the same time, the technical standard-setting ability—that is, the ability of the organization to successfully manage the development process of technical standards—is the key to standardization. Finally, the standard-promotion ability is the guarantee that organizations will achieve their strategic objectives and reap economic benefits by using the standards. Based on the unique characteristics of organizations involved in standardization, these three key indicators actually better reflect the level of standardization that nonprofit organizations can achieve. The more the conditions can be met, the higher the level of standardization. The more nonprofit organizations possess a high degree of standardization capability, the greater a country's overall development of association standardization.

China's government-backed national institutes, associations, chambers of commerce, unions, and industrial technology alliances have a high level of standardization because these organizations receive generous support from the government, including funds, personnel, and policies. For example, the China Communications

Standards Association (CCSA), established with government support, can easily obtain resources under government control. As the association is commissioned by governmental departments to develop national and industry standards for communications technology, it can also get subsidies for standardization from the government. Since its establishment in 2002, the CCSA has developed 367 national standards and 3,022 industry standards, which cover both traditional and new areas of communications.

However, China still lacks an equivalent to the multitude of consortia composed of large enterprises that are found in the United States. In terms of standards-based technology R&D and innovation capacity, this is a weakness of Chinese enterprises. They have a long way to go before they catch up with the United States.

In the United States, many standardization organizations (such as the ASTM, ASME, and IEEE) produce standards that are widely accepted. Standards set by these organizations are adopted across the globe, and are well regarded in the industry and the field of standardization in China. Their influence is even greater than ISO standards. Many Chinese industrial exports to the United States have adopted the standards set by these organizations (Wang and Liang 2013). In addition to professional standardization bodies, corporate interest groups (i.e., consortia) formed by US companies, both large and small, also have strong standardization capabilities, especially in the field of information and communications technology (ICT). Organizations such as the World Wide Web Consortium (W3C) and the Internet Engineering Task Force (IETF) demonstrate the highest innovation capability in technology development today.

> *China still lacks the multitude of consortia composed of large enterprises that are found in the United States*

Returning to standardization organizations, the American Society for Testing and Materials (ASTM) serves as an illuminating example. The ASTM is one of the oldest and largest nonprofit academic organizations in the United States. More than a century following its establishment in 1898, the ASTM has 33,669 individual and group members, organizes more than 35,000 experts working on various technical committees, and has developed more than 12,000 standards. The ASTM has a total of 132 technical committees mainly responsible for the development of performance standards, test methods, and procedures in the fields of materials, products, systems, and services. Although the ASTM standards are developed by a nonofficial professional education body, it has won official trust from American industries because of the high quality and ease of adaptability of its standards. The standards have been adopted not only by the private sector, but also by the US Department of Defense (DoD) and various federal agencies. In the past 25 years, the DoD has worked with the ASTM to replace the US military standards with voluntary standards. Currently, there are more than 500 people in the DoD actively participating in the activities of the ASTM. So far, 2,800 US military standards have been replaced by ASTM standards. As reforms to the DoD acquisition system progress, the US military is sure to adopt ASTM standards more often. In addition to the DoD, other federal agencies are also using a number of ASTM standards, and have established a broad and close cooperative relationship with the ASTM. The wide adoption of ASTM's standards validates the criteria proposed by Zeng Deming: technical superiority, technical standard-setting ability, and standard-promotion ability. In each area, the ASTM excels.

> *The American Society for Testing and Materials has developed more than 12,000 standards*

Government attitudes toward standardization bodies

Government attitudes are an important external factor that influences the development of association standardization. When a government supports an activity through policy—whether industrial, economic, or social—it will quickly grow. Numerous examples confirm this, from the rapid rise of Japan after World War II, to the economic development of Southeast Asian countries in the last century, to China's economic achievements. Andrei Shleifer and Robert W. Vishny (1998) put forward three perspectives on how to view the government. The first is "the invisible hand": good market operation requires the government to create basic functions necessary to make it work, such as the provision of law, order, and defense. The second is "the supporting hand": free markets come with many problems and the government should intervene in the economy to correct market failures and maximize social welfare. (The third one is "the grabbing hand": the goals of both authoritarian and democratic government are not to pursue the maximization of social welfare, but to achieve their own interests.) In the case of association standardization, government will use both the "invisible hand" and the "supporting hand" to promote an environment favorable to the rapid development of association standards. It can do this by developing favorable macro policies that support the creation of standards, improving relevant laws and regulations, maintaining a good market environment and fair competition, and providing comprehensive public services.

The Chinese government is now supporting and encouraging the development of association standards through laws and policies. Legally, the new amendment to the Standardization Law (revised draft) adds provisions for association standards for the first time. The law specifically states that "social organizations established lawfully can set association standards. The setting of association standards shall be standardized, guided, and supervised by the competent administrative department of standardization under the State Council." This is one of the highlights in the amendment to the Standardization Law, marking a clear legal status for association standards in China's standard system.

In terms of policies, the relevant documents issued by the Chinese government clearly support the development of association standards. On March 1, 2016, the General Administration of Quality Supervision, Inspection, and Quarantine and the Standardization Administration of China (SAC) jointly issued the *Guiding Opinions on the Cultivation and Development of Association Standardization* (hereinafter referred to as the *Guiding Opinions*), a programmatic document for the development of association standardization in China. The *Guiding Opinions* identifies the main objectives of the cultivation and development of association standards:

- By 2020, association standards set by the market will have been better developed, thus better satisfying the needs of market competition and innovation.

- The number and competitiveness of association standards will have steadily increased.
- Association standard-setting bodies will possess significantly more influence.
- The association standardization mechanism will attain a level of sophistication.

The *Guiding Opinions* also points out that association standards will be set, selected, and adopted by the market voluntarily. In the absence of national standards, industry standards, and local standards, social associations may develop association standards in response to innovations and market demand, thus filling the voids. Social associations are encouraged to develop standards that are stricter than national standards and industry standards so as to promote the development of industries and enterprises and to enhance the market competitiveness of products and services. Further, association standards should compete with each other in the market, using market mechanisms, thus increasing their quality, popularity, and likelihood of adoption.

The *Guiding Opinions* stresses that it is necessary to:

- Establish a mechanism for transforming association standards into national standards, industry standards, and local standards.
- Specify the necessary conditions and procedures for the transformation.
- Encourage association standards that (1) pass a "good behavior" evaluation; (2) have a positive effect on an industry; (3) cover subjects that would normally be within the scope of the national standards, industry standards, and local standards, thus facilitating their transformation into equivalent national standards, industry standards, or local standards.
- Clear the way for social associations to participate in international standardization activities and encourage social associations to put forward international standard proposals and participate in their drafting.

The *Guiding Opinions* states that competent departments of standardization under the State Council shall establish an information platform to strengthen transparency and the ability of society to monitor standards. Each provincial administrative department of standardization may tailor their own association-standard information platforms, which shall be connected with the national platform.

The laws of the United States protect and promote the operation and development of voluntary standards in many aspects. The National Technology Transfer and Advancement Act (NTTAA), approved by the United States Congress in February 1996, requires federal agencies to use voluntary standards instead of government-unique standards wherever possible. To ensure the effective implementation of the NTTAA and the adoption of voluntary standards as a long-term policy, shortly after the release of the NTTAA, the US Office of Management and Budget issued *Circular No. A-119—Federal Register (Federal Participation in the Development and Use of Voluntary Consensus Standards and in Conformity Assessment Activities)* (hereinafter referred to as *Circular No. A-119*), which was revised in 1998 and again in 2016. *Circular No. A-119* is a supplementary document to ensure the implementation of the NTTAA. In August 2004, the United States issued the Standards Setting Organizations Promotion Act, which mainly protects the rights of voluntary standard-setting organizations and is designed to encourage them to develop standards. In addition, the United States Congress passed the Telecommunications Act and the US Consumer Product Safety Act to put forward specific requirements for the adoption of voluntary standards by the federal government.

Federal US agencies are required to use voluntary standards instead of government-unique standards wherever possible

In comparing how effectively China and the United States back their association standards, it is clear that the United States strongly supports voluntary standards through sophisticated laws and systems. In contrast, due to the immaturity of its association standards, China has only released a few encouraging policies and indefinite incentives for association standards, offering support from the macro level but lacking specific operational instructions. In China, association standards are not referred to in the national laws, regulations, and standards. In addition, there are few government officials participating in the standardization activities of nongovernmental organizations and giving support to association standards.

Market demand

The primary driver behind the creation of association standardization is correctly understanding the needs of the market and the demand for standardization. Association standards, regardless of their ability to foster self-discipline and technological advantages, are essentially designed to ensure that their creators keep up with the market. For this reason, we can consider a standard as a special "product" designed to meet special needs. Useful standards can only come from market demand and from an accurate understanding and careful judgment of issues pertaining to technology, management, and service in the market economy (Wang 2014).

As major players in market-based economic activities, enterprises, consumers, and the government can all generate strong demands for association standards. For enterprises, incorporating their own mature technologies in association standards can help disseminate their production techniques, bring direct gains in increased revenue, and lift their competitiveness. For consumers, association standards can offer more defined classifications that ensure product comfort, economic efficiency, and technological quality. This enhances a product's credibility and reliability in fully meeting consumers' demands, as manufacturers have access to more specific market requirements and testing methods to ensure conformance to the standard. For governments, association standards can promote technological progress and more quickly eliminate inferior products, with the final result of upgrades to the entire industrial sector. Additionally, by stimulating comprehensive international exchanges, as well as increasing the speed and quality of those exchanges, association standards break trade barriers set by other countries, cut down R&D and manufacturing costs, expand the range of international trade, and, finally, galvanize international trade. In other words, association standards can strengthen the foundation for the orderly development of the market and facilitate industrial upgrades and international trade development.

> *Useful standards can only come from market demand*

In China, tremendous market demands have helped to speed up the development of association standards. Take the standard set by the China Solid State Lighting Alliance (CAS) as an example. The CAS was founded in October 2004 by 46 domestic enterprises, universities, and research institutions in the semiconductor lighting industry on the principle of "voluntariness, equality, and cooperation." The CAS now boasts more than 530 members, with a combined output accounting for over 70 percent of the industry in China. Among the members are more than 20 listed enterprises, including the top four traditional lighting enterprises, the Chinese branches of the top five international enterprises, the top five Taiwanese enterprises, and the top ten testing institutions of China. In 2015, the Chinese semiconductor lighting industry totaled 424.5 billion yuan in worth.[2] Over the past 10-plus years, China has gradually narrowed its gap with international competitors in the semiconductor lighting sector. The industry has matured and the industrial chain has continued to improve.

The LED industry is a different story, however. The LED lighting field features a low access threshold and simple assembly process, leading to large differences in product quality levels and a wide gap in product prices. For example, prices of bulbs of the same type vary from five yuan to fifty yuan. The main reasons for such market disorder and difference in product quality lie in the immaturity of standards, testing, and certification systems. Rapid development of the LED industry led to calls for relevant product standards and technical specifications. Unfortunately, technical committees and industry-standard bodies were then established on a traditional top-down basis and administered by different governmental departments. As a result, they were unable to respond in a timely way to rapid change. Meanwhile, the standard-setting procedures emerged relatively slowly and failed to promptly reflect the changes in market demand and industrial development, leading to calls for a new standardization mechanism.

Although semiconductor lighting (which has the benefit of standardization) has entered the field of functional lighting, such as road lighting, it will be some time before a relevant national standard for LEDs is released. This is because the technology is less mature at present and no ready testing methods or standards are available for reference either at home or abroad. The LED industry urgently needs relevant standards and specifications to be released, so as to guarantee the development of LED functional lighting products and encourage energy-saving choices. To this end, the CAS took the initiative to advance LED streetlight standardization and released the alliance's first technical specification, "Integrated LED Streetlight Measurement Method," in 2008. Some technical provisions in the standard have been recognized and adopted by the energy star standard of the US Department of Energy. So far, the alliance has formulated 36 standards covering LED product performance, measurement methods, power supply, and application interface (Wang and Liang 2016).

Standardization in the United States was first driven by the need for technical compatibility. Voluntary standard-setting organizations initially emerged in the private sector to solve production and supply chain problems, and industrial associations formulated technical standards for their own respective fields. As science and technology developed more rapidly, the life cycle of technologies and products in the ICT field in particular became shorter and shorter. Technological upgrades outstripped traditional standards bodies. Enterprises in fields with fast technological advances began to realize the importance of setting new technical specifications as early as possible to develop new markets. However, if standard-setting procedures followed the official line completely, the process would obviously be too long and time-consuming. A single enterprise without considerable clout in the market would be unable to control standards on its own. In such a situation, enterprises with common market interests would need to compromise with each other and form an association to jointly create association standards, thereby meeting the demands of quickly changing technologies and industries.

Enterprises in fields with fast technological advances realized the importance of setting new technical specifications as early as possible

Foreign experiences

Developing countries, with limited abilities and experiences, always attach high importance to the successful experiences and practices of developed countries. This influence is known as benchmarking. Benchmark management is not only an effective approach common in business management, but has also become a systematic tool to continuously improve competitiveness in many countries, as governments increasingly value the international strength of their industries, enterprises, and countries (Kong and Cheng 2004). Association standards have a history spanning one hundred years in the United States, which has allowed them to become very sophisticated and instrumental in achieving economic and social benefits. These benefits are undoubtedly very attractive to China, given its ambition to cultivate association standards.

At present, there are two major international standardization management models: the US model and the EU model. The decentralized, self-governing voluntary standard management model of United States is different from that of the EU and other countries internationally. In fact, the management model is wholly unique. It is closely tied to the US economic system, its development level, economic growth model, history, and culture. As China's economy has progressed since the reform and opening up, many of its practices have drawn on the experiences of developed Western countries. An important reference for China's standardization reform, particularly its association standards reform, is the US government's experience in voluntary standards, while its compulsory standards reform mainly refers to the EU's technical regulations management model.

Fast-tracking Association Standardization in China

How to develop association standards?

In the newly revised Standardization Law, association standards have been included in the national standard system, and relevant policy documents have been released to encourage the development of association standards. These are significant acts on China's part, but they are not nearly enough. Further specific policy measures are imperative to achieve a number of goals.

1. To provide nongovernmental standardization organizations with a liberal environment: Nongovernmental standardization associations, alliances, and other organizations are a major power driving the development of association standards in China. The government should carry out reforms to make it easier for private enterprises

Table 1. Comparison of Factors Influencing Association Standardization Development in China and the United States

	China	United States
Nonprofit organization development level	Nongovernmental and non-profit organizations are still fledging; many societies, associations, commercial chambers, and federations have government backgrounds, and standardization organizations have a short history.	Nongovernmental and non-profit organizations are mature, and standardization organizations have a long history.
Nonprofit organization	Few government-background national societies, associations, commercial chambers, federations, or industrial technical alliances have high standardization levels, but most nonprofit organizations have poor standardization ability.	Many standardization organizations enjoy high international reputation, and their standards are greatly influential internationally.
Attitude of government	Government encourages association standards development, but relevant laws, regulations, and policies lag behind, and no specific policy measures are available.	Government strongly advocates use of voluntary standards, has rolled out relevant law and policy documents, and has formulated operable policy measures.
Market demand	Standards are focused on developing market order and acquiring competitive advantages via technology applications.	Standards are designed to meet the demands of quickly changing technologies and industries.
Foreign experience	Looks to the US voluntary standard management experience.	N/A

to jointly establish nonprofit institutions and, in particular, to ease the limitations on nongovernmental parties establishing industry standardization associations. This will create a good legal environment for the rise and healthy development of nongovernmental standardization groups. The government should also withdraw from the control of industrial recommendatory standards, and leave the setting of vital product standards completely to standardization associations.

2. To make membership to foreign enterprises and institutions less restrictive: At present, some association standardization organizations in China, including some industrial associations and alliance organizations, invite the participation of foreign companies. For example, the China Communications Standards Association (CCSA) has foreign observers that are entitled to participate in the CCSA's business meetings, including members' conferences and activities; to submit documents in their branches and technical committees; to obtain phasic work documents and published documents of standards; to acquire the association's public journals and technical information; and to criticize, make proposals for, and supervise the association's work. Some American enterprises, such as Intel, Qualcomm, IBM, Apple, Oracle, and Cisco, have participated in CCSA's standard-setting activities as observers. In another example, the China Solid State Lighting Alliance (CSA) includes multinationals such as Philips and Osram as member units. But China's alliances have not opened up enough to foreign enterprises. In order to increase the capability and influence of association standardization organizations, limitations on foreign enterprises and institutions should be further loosened in the future to allow more to become official members and to participate in domestic association standardization activities.

> *Limitations on foreign enterprises and institutions should be further loosened in China*

3. To encourage reference to association standards in laws and regulations: The legislature should consider whether there are current association standards applicable when making laws and regulations that involve specific technical requirements. If there are, these standards can be referred to, and their relevant contents can be included in the laws and regulations. If not, the legislature can authorize its trusted standards bodies to create them. (The legislature should dispatch its representatives to participate in the standard-setting process, so as to ensure that their opinions are reflected in the standards.) The standards thus created should be used in or referenced by relevant technical regulations. This practice will

greatly lift the status of association standards and promote the development of the standards.

4. To gradually remove the limitations on the qualification of association standard-setting bodies: The latest revised Standardization Law stipulates that all social organizations established according to law can set association standards, and that their setting shall be standardized, guided, and supervised by the competent administrative department for standardization under the State Council. In the current experimental projects involving association standardization, all pilot associations are required to have a legal personality. Such provisions on the qualification of standard setters have both positive and negative effects.

First, the positive effects. Around the world, international organizations such as the International Organization for Standardization (ISO) and the International Electrotechnical Commission (IEC), as well as developed countries such as the United Kingdom and Canada, emphasize that standard-developing organizations should be legal entities or administrative entities that can take legal responsibility for all their standard-developing activities. Domestically, China's current standard system also requires standard-setting institutions to be legal entities or administrative entities. The major setters of national standards, industrial standards, and local standards are administrative entities sponsored by the government at all levels. Setters of corporate standards, in contrast, should be business entities. In terms of China's administration of social organizations, Article 3 of the *Regulations for Registration and Management of Social Organizations* clearly requires that "social organizations should be equipped with legal personality."

To summarize, as a new kind of standard, association standard setters should have legal personality, thus allowing them to undertake legal responsibilities for all their standard-setting activities and to prevent confusion with corporate standard setters, as well as possible chaotic situations. In terms of alliances that have obtained a legal personality, the standards they set all belong to the realm of association standards; if they do not have a legal personality, they are corporate standards, and legal persons must assume the relevant responsibilities for the corporate standards.

Next, the negative effects. The Standardization Law limits association standard setters to social organizations with legal personality, which has a highly adverse influence on the development of association standards. First, association standardization is a bottom-up activity. An association standard is created to meet market needs for that standard. Therefore, any organization and any institution should be able to formulate association standards and make the standards applicable in the particular market segment that requested it, with no government intervention.

Taking the United States as an example, any social organization is allowed to set its own association standards, and any such standards are available for society to use voluntarily. For one standardization process, then, there might be multiple applicable standards set by different social organizations. This bolsters competition and ensures that the best standards are more likely to be chosen. The whole process is based completely on market behavior. Competition both encourages the survival of the fittest and weeds out poorer association standards. The competitive environment pushes social organizations to work hard to set standards that meet users' needs and are acceptable to society.

Second, setting limitations on who qualifies to be a standard setter demonstrates an interventionist government mindset. Currently, "streamlining administration and delegating more powers to lower-level governments" is considered to be the main task of reform. But in actual practice, many government departments still hold the power to intervene in the economic activities of market entities. This will have a chilling effect on social organizations, and will undoubtedly keep a large segment locked out from setting association standards due to their lack of the legal personality required by the government. This situation

Setting limitations on who qualifies to be a standard setter demonstrates an interventionist government mindset

is unfavorable for the development and expansion of association standards. Currently, China's social organizations face multiple roadblocks, including the categorized-administration principle and the noncompetition principle, both of which are related to the traditional administrative system of a planned economy. The system sets a threshold that common citizens can never pass and that largely prevents nonprofit organizations from gaining legal status through registration (Wang 2006). It is clear that the current system's requirement that nonprofit organizations become legal entities is too strict, and the procedures to be registered are too complicated (Yu and Li 2009). All such practices are unfavorable to the development of social organizations in China.

Third, limiting the number of qualified association standard setters will certainly cause insufficient competition. The principle of "independent development, free selection, and voluntary adoption" is how association standards are envisioned in the *Guiding Opinions*. However, if capable social organizations are excluded because of barriers to achieving the relevant qualifications, the principle becomes highly compromised in reality. In fact, the lack of qualified organizations limits competition.

During this initial period of association standardization reform, China can maintain the practice of letting only social organizations with legal personality formulate association standards. But the practice should be an interim measure rather than a permanent policy. In the future, as association standards become more mature and standard-setting organizations more regulated, the qualifying limitations on association standard setters should be removed, allowing all social organizations to set standards as long as they have the capabilities.

Social organizations should be allowed to develop association standards even when there are available national standards

5. To expand the scope of association standard setting step by step: The Chinese government should gradually cancel the provisions in the *Guiding Opinions* that "social organizations can develop association standards where no national standard, industrial standard, and local standard are available, so as to quickly respond to the demands of innovation and the market and fill in the vacuum." In their place, social organizations should be allowed to develop association standards even when national, industrial, and

local standards are available. Existing industrial standards can be adapted into national standards or, failing that, adopted by corresponding social organizations. China's certification and accreditation systems and testing institutions should thoroughly change their current practices of not recognizing association standards.

What will China's model look like in the future?

Association standardization in China should be guided by the government, led by social organizations, and supported by technical organizations. Recommendations for how this can be done are described below.

1. **Government, the guide:** Government departments should play the role of guide and coordinator in association standards development. Departments can offer opinions on association standards development, as well as regulate the formulation, revision, and registration of association standards; draw up a code of conduct and take advantage of third-party evaluation and credibility mechanisms of social associations undertaking standardization activities; carry out pilot projects to advance social association standardization; and provide incentives to encourage enterprises to participate in formulating and implementing association standards.

2. **Social organizations, the main players:** Industrial associations and other social organizations are important bridges for communication and contact among governments, enterprises, and relevant technical institutions. These social organizations can familiarize enterprises with government policies and convey the needs and wants of enterprises to the government. They can coordinate the relations and interests of all related enterprises and facilitate cooperation among research institutions, testing institutions, and enterprises. Therefore, industrial associations and other social organizations are core forces for organizing the setting of association standards, and they are major responsible parties for the creation of association standards. These organizations need to be given a leading role in setting and implementing association standards.

3. **Technical organizations, the supporter:** Colleges, universities, research institutions, and testing institutions are resources that have advanced testing equipment, abundant testing resources, and insight into products and industrial trends. Because of these resources, these technically capable organizations should actively participate in the formulation of association standards, with the goal of ensuring that the final standards have advanced technical indicators and high operability. The technical institutions for standardization should also strictly follow national laws and regulations and the requirements of compulsory standards, maintain high-level quality control, and provide enterprises with relevant consulting services.

What can China learn from the US system?

By studying the American voluntary standard system, China can understand its merits and limitations, and its applicability to their own reform efforts. The American voluntary standard system comes from a solid economic and technical basis, strong marketing ability, and a cultural tradition that advocates freedom and democracy. The market-oriented, decentralized, self-governing voluntary standard system reflects the demands of different interested parties, and is an important source for US industrial innovation. The strong points of this voluntary standard system are the following:

- First, decentralized self-governance protects the rights of social organizations to set standards. The American standard system is characterized by a high level of openness and voluntary participation, and standards are set by groups in the form of technical committees (Li 2004).
- Second, the standard system is driven by market mechanisms to meet users' demands. The relationship between the American National Standards Institute (ANSI) and associations and alliances is established through recognition and examination mechanisms. Free from administrative subordination, behavior and operations are standardized by constitutions, deeds, agreements, and procedures.
- Third, there is public-private collaboration in setting standards. In essence, the most typical characteristic of the American standard

system is that "private and public sectors set standards collabora-tively" (Russell 2006). At least on principle, such multilateral co-operation can prompt different stakeholders such as governments, industries, and customers to ex-press their demands and increase the likelihood that a standard will be adopted. For example, when the American smart grid was upgraded, the federal gov-ernment, through public-private

> *The most typical characteristic of the American system is that private and public sectors set standards collaboratively*

partnerships, was able to accelerate the speed at which standards were set and that improvements to the grid were made.

- Fourth, the standard system embodies the principle of "who uses will be benefited." The government does not need to channel funds into standard setting. To survive, all standardization organizations depend on membership fees and from the income generated from sales of standards documents. Therefore, standardization organi-zations focus their efforts on attracting more members, in part by creating standards that are widely recognized and respected. It is obvious that the mechanism effectively allocates resources via the market economy (Xu 2001).

While the American voluntary standard system boasts many strong points, some deficiencies have existed since its establishment, mainly in the following aspects:

- First, there are fierce conflicts among standard-setting parties and a disregard for the public interest. The number of entities par-ticipating in standardization increases the amount of conflict and competition, which can negatively impact the effectiveness and fairness of the American standard system. For example, the dis-pute over the open file standard (such as Microsoft's OOXML and ODF standards) made the American standard system less efficient (Cargill 1989).[3] In addition, in the fragmented, market-driven voluntary standard system, profit-oriented enterprises often lead the standardization process. This can steer the process away from the goal of supporting public policy, especially when it comes to important strategic standards such as smart grid standards. Such standards are very important for driving innovation and building national capacity for further innovation. Setting and maintaining important strategic standards needs massive human, material, and financial resources. If private interests occupy an important position in the standardization process, the public interest will inevitably be overlooked (Ernst 2013).

- Second, the weak role of the US government and a lack of effective coordination limit the system's effectiveness. The United States has always believed in the role of the market to drive society and the economy, and its government rarely intervenes. In reality, it is a major weakness of the American standard system that the government plays such an insignificant role. Standards are competitive and continually changing, a characteristic that is particularly striking in basic technical fields. Therefore, the government should not only be a standard user, but should also play the role of standardization activity supervisor and coordinator. However, that is exactly where the US government fails. Another weak point of the American voluntary standard system is that there is no effective coordination among hundreds of competitive private standard-setting organizations NIST is not charged with any role in the making of the US standard system. While it is a governmentally charged organization and can create standards, private sector protests have largely prevented it from assuming any greater role in US standardization. Likewise, the American National Standards Institute's (ANSI) role is limited, despite acting as a coordinator in the voluntary standard system. The ANSI is a private organization, and thus unable to reduce competition and conflicts among other private organizations. There is no catalogue of US standardization organizations (SSO and SDO), a situation excerbated by the fact that many consortiums do not believe that they are US organizations, but rather are international in scope.
- Third, abuse of patents is a problem. With technology competition intense, any enterprise that wants a competitive advantage must have the "essential patents" of standards, which are of great importance. K. Blind et al. (2004) called essential patents a strategic weapon, and one that can have significant influence on the standardization process. Such a patent strategy makes it very hard

to implement the American standard system's principle of "public availability." American standards are driven mainly by enterprises holding many essential patents, which points to a fundamental weakness of the American standard system: standard users, including executors and particularly end users, have no say in the system.

- Fourth, the concept of open standards is obscure. Due to lack of effective coordination, another major defect of the American standard system is the undefined concept of open standards. Open standards have become a common belief in the American standard system, but "all suppliers just talk about open systems in words and have not reached any unanimous agreement in deed" (Libicki et al. 2000). A case in point is the concept of "voluntary consensus standards." The system of voluntary consensus standards is at the core of the American standard system. The National Technology Transfer and Advancement Act (NTTAA) defines them as "the standards set through certain processes, which include the five fundamental principles of openness, transparency, interest balance, due process/appeal process, and consensus" (OMB Circular A-119, 1998).[4] But, first, the definition fails to set a boundary for the voluntary consensus standards. It clearly allows other private sectors' standards to include "non-consensus standards," "industry standards," or "de facto standards," which is not consistent with the characteristics of openness as defined by the US Office of Management and Budget. Secondly, it does not clearly specify the role of alliances, which also leads to an inconsistency in how the regulations are implemented. For example, government organizations must report to the National Institute of Standards and Technology (NIST) about their adoption of voluntary consensus standards, while alliances are free from these requirements (Garcia et al. 2005). The above-mentioned problems show that the concept of open standards in the American standard system is obscure and hard to implement in practice.

Another major defect of the American standard system is the undefined concept of open standards

Therefore, when advancing comprehensive reform of the standardization system, China needs to draw on the experiences of the American standard system to adapt the best aspects, while avoiding the flaws.

- First, wholesale duplication of the American standard system would have obvious limitations (Ernst 2013). A country's national economic system, development level, economic growth model, and historical and cultural uniqueness will influence its standard system. When facing similar challenges in standardization, countries differ notably from each other.
- Second, China should attach full importance to the role of non-governmental standardization organizations. In particular, China should rely on societies (associations) and other social organizations to establish the standard-setting system step by step, with enterprises as major players and associations as the core. The country should form a standardization mechanism that is suitable for the development of a market economy, while maintaining self-discipline and order.
- Third, the government's role in supervising and coordinating efforts should be stressed. In the American voluntary standard system, the government's role is too weak, increasing conflicts and lack of effective coordination among all stakeholders, which detracts from an effective, open standardization process. Therefore, China should, on the one hand, give play to the decisive role of the market in allocating standardization resources. On the other hand, it should pay close attention to the government's role of macro administration and coordination in establishing a fair competitive environment and providing public services for standardization and other aspects.

Balancing IPR systems and association standards worldwide

The differences between developed and developing countries in terms of legal systems, jurisdictional authority, technological progress, and cultures have caused a long-term imbalance in the development of intellectual property rights (IPRs). Strictly litigated and enforced IPR regimes ensure that developed countries are able to transform their absolute technological advantages into economic benefits. This, in turn, reinforces their monopoly in the global high-tech field.

Meanwhile, developing countries, hampered by their comparatively limited knowledge base and R&D ability, prefer to utilize relatively loose, moderate IPR-protection mechanisms. This allows developing countries to make full use of the international diffusion of technology and transfer of knowledge, and helps them accelerate development of their industrial technologies. Therefore, it is neither necessary nor wise for developing countries to offer IPR protection to the same extent that developed countries do. While the optimal IPR protection suitable for developed countries is higher than developing countries, overprotection of IPR will hinder the course of developing countries' technological development (Wang 2011).

Developing countries prefer to utilize relatively loose, moderate IPR-protection mechanisms

Under certain conditions, IPR protection can promote a country's technological innovation, but when that protection is too strong, innovation can be dampened or even stopped. Therefore, an optimal degree of IPR protection needs to be established. The design of an IPR system should balance two key factors: on the one hand, IPR protection cannot be too low, thus ensuring knowledge "exclusiveness" and return on innovation, and motivating researchers and developers to innovate. On the other hand, IPR protection cannot be too high, thus guaranteeing that patent owners are not able to monopolize an industry and cause market distortion and an out-of-balance allocation of resources.

Endnotes

1. Source: http://nccs.urban.org/statistics/quickfacts.cfm.

2. Source: http://www.china-led.net/news/201512/31/31868.html.

3. OOXML means Microsoft-developed open office XML document format, while ODF standard means a document format developed by the Organization for the Advancement of Structured Information Standards (OASIS), started by IBM, Sun Microsystems, Oracle, etc.

4. OMB Circular A-119 means the OMB Circular A-119 on Federal Participation in the Development and Use of Voluntary Consensus Standards and in Conformity Assessment Activities (OMB Circular A-119 for short).

Bibliography

Alic, John. 2009. "Energy Innovation from the Bottom-Up: Project Background Paper." Paper prepared for the joint project of the Consortium for Science, Policy, and Outcomes (CSPO), Arizona State University, and the Clean Air Task Force (CATF), March.

Aoki, Masahiko, and Masahiro Okuno. 2002. *The Role of the Market, the Role of the State*. China Development Press.

Blind, K., et al. 2004. *Interaction between Standardization and Intellectual Property Rights*. Technical Report EUR 21074 EN, European Commission Joint Research Centre, 248.

Cargill, C.F. 1989. *Information Technology Standardization: Theory, Process, and Organizations*. Bedford, MA: Digital Press.

China National Institute of Standardization. 2007. *Study of Several Major Theoretical Issues of Standardization*. Standards Press of China.

Criqui, Francis L. "Tex." 2004. "On Four Major Elements of Competitiveness from the American Standard System." *Shanghai Standardization Monthly* (7): 15–17.

Ernst, Dieter. 2013. *America's Voluntary Standards System: A Best Practice Model for Asian Innovation Policies?* Honolulu: East-West Center, Policy Studies No. 66.

Garcia, D.L., B.L. Leickly, and S. Wiley. 2005. "Public and Private Interests in Standard Setting: Conflict or Convergence?" In *The Standards Edge: Future Generations,* edited by Sherrie Bolin, 126–130. Ann Arbor, MI: The Bolin Group.

Gilpin, Robert. 2006. *Global Political Economy: Understanding the International Economic Order*. Shanghai: Shanghai People's Publishing House.

Hart, D.H. 1998. *Forged Consensus: Science, Technology, and Economic Policy in the United States, 1921–1953.* Princeton, NJ: Princeton University Press.

He Ming. 2014. "Contrast Analysis of Institution Standards of China and the US." *Standardization of Engineering Construction* (10): 52–56.

Holland, Thomas P. and Roger A. Ritvo. 2008. *Non-profit Organizations: Principles and Practices.* New York, NY: Columbia University Press.

Kang Junsheng and Yan Shaoqing. 2015. Analysis and Thought on Development of Social Group Standards. *Standard Science* (3): 6–9.

Kirsh, B.S., and H.R. Shapiro. 1939. *Trade Associations in Law and Business.* New York, NY: Central Book Company, 135–163.

Kong Jie and Cheng Zhaihua. 2004. "Review of Benchmark Management Theory." *Journal of Dongbei University of Finance and Economics* (2): 3–7.

Li Fengyun. 2004. "Investigation Report of America's Standardization." *Metallurgy Standardization and Quality* 42 (4): 55–61.

Liao Li and Cheng Hong. 2013. "Study of Law and Standard Consistency Model: Based on the Perspectives of Hard Law and Soft Law and China's Practice." *China Soft Science* (7): 164–176.

Libicki, M., J. Schneider, D.R. Frelinger, and A. Slomovic. 2000. *Scaffolding the Web: Standards and Standards Policy for the Digital Economy.* Santa Monica, CA: RAND Science and Technology Policy Institute.

Liu Fei. 2009. "The Value of Standards Is Rooted in System and Strategy: Comparison of Standardization Systems and Strategies Between China and the US." *Communications Standardization* (11): 20–22.

Liu Jin and Wang Yanlin. 2012. "Study on Association Standards and Standardization Law." *Wuhan University Journal* 65 (3): 90–93.

Liu Qingchun. 2012. *General Introduction to Standardization in the US, the UK, Germany, Japan, and Russia.* China Zhijian Publishing House, Standards Press of China.

Liu Sanjiang. 2016. "On the Basic Logic of China's Standard System Reform: From the Perspective of Standard Interest Competition." Working paper.

Liu Sanjiang and Liu Hui. 2015. "The Thought and Path of China's Standardization System Reform." *China Soft Science* (7): 1–12.

Lundqvist, Björn. 2014. *Standardization Under EU Competition Rules and US Antitrust Laws: The Rise and Limits of Self-regulation.* UK: Edward Elgar Publishing, 149–183.

Mattli, W. and T. Buethe. 2003. "Setting International Standards: Technological Rationality or Primacy of Power?" *World Politics* 56 (October): 1–42.

OMB Circular A-119, 1998. "Federal Participation in the Development and Use of Voluntary Consensus Standards and in Conformity Assessment Activities" https://obamawhitehouse.archives.gov/omb/circulars_a119/

Pelkmans, Jacques. 1987. "The New Approach to Technical Harmonization and Standardization." *Journal of Common Market Studies* 25 (3): 249–269.

Robert, R.J. 1999. *Governing in the Absence of Government: The Birth and Development of the United States Industrial Standards System.* PhD dissertation, University of California, Santa Barbara.

Russell, Andrew L. 2006. "Industrial Legislatures: The American System of Standardization." In *International Standardization as a Strategic Tool.* Geneva: International Electrotechnical Commission (IEC), 70–79.

Salamon, Lester M., et al. 1999. *Global Civil Society: Dimensions of the Nonprofit Sector.* Baltimore, MD: The Johns Hopkins Center for Civil Society Studies.

Shang Huimin. 2014. "Current Situations of Guangdong Industrial Cluster Alliance Standard Development and Study of the Countermeasures." *Guangdong Science and Technology* (7): 3–5.

Shleifer, A. and R. Vishny. 1998. *The Grabbing Hand: Government Pathologies and Their Cures.* Cambridge, MA: Harvard University Press.

Song, Hualin. 2009. "The Establishment and Evolution of the Legal System of Technical Standards in Contemporary China." *Study and Exploration* (5): 15–19.

Updegrove, Andrew. 2010. "When and How to Launch a Standards Consortium." *Standards Today* 9 (3). http://www.consortiuminfo.org/bulletins/sep10.php#feature.

Wang Hua. 2011. "Is More Stern IPRs Protection System Favorable for Technology Innovation?" *Economic Research Journal* (2): 124–135.

Wang Ming. 2006. "The Social Function and Classification of Non-Profit Organizations." *Academic Monthly* (9): 8–11.

Wang Ming and Jia Xijin. 2003. "China's Non-profit Organization: Definition, Development and Policy Suggestions." In *Party III NGO and China, Social Transition and Non-governmental Organizations under Globalization,* compiled by Fan Lizhu, 263–264. Shanghai: Shanghai People's Publishing House.

Wang Ping and Liang Zheng. 2013. "Research on the Standardization Development of Chinese Association and Alliance." *China Standardization* (8): 59–62.

Wang Ping and Liang Zheng. 2016. "Current Situations of China's Non-profit Standardization Organization Development: Case Study Based on Organization Characteristics." Working paper.

Wang Xia and Lu Lili. 2010. "Study on Associational Standardization in China." *Standard Science* (4): 29–32.

Wang Zhongmin. 2014. "Where Are Valuable Standards From?" *China Standardization* (1): 43–45.

Xu Jingyue. 2001. "Review on America's Standardization System." *China Standardization* (4): 48.

Yao Yuan and Song Wei. 2010. "Comparative Study on the Formation Pattern of Patent Pools under the Trend of Patent Standardization—DVD Model Vs MPEG Model." *Studies in Science of Science* 28 (11): 1,683–1,690.

Yu Xiang and Li Na. 2009. "Comparison and Browning Ideas from Organization States of Non-profit Organizations at Home and Abroad." *Journal of North China Electric Power University (Social Sciences)* (5): 41–45.

Zhang Shuqing. 2007. "American National Standard Management System and the Operation Mechanism." *World Standardization & Quality Management* (10): 17–19.

Zeng Deming, Wu Yanwu, and Wu Wenhua. 2005. "Study on Constructing the Index System of Enterprise Technical Standardization Ability." *Science and Technology Management Research* (8): 164–167.

www.ingramcontent.com/pod-product-compliance
Lightning Source LLC
Chambersburg PA
CBHW071342290326
41933CB00040B/2092